Overcoming
DENTAL
ANXIETY

By: Dr. Chi Mba

COPYRIGHT NOTICE

TABLE OF CONTENTS

ACKNOWLEDGEMENTS

I am so grateful to God who has allowed me this incredible opportunity to be a part of this profession that changes people's lives in such a profound way. I will like to thank my father, PJ Nwachukwu, my 5 beautiful children, Ada, Mma, Ugo, Ola and Uzo who always believe in me. Thanks to my husband, Matthew who always brings out the strength I never knew I had. Special thanks to all my team and my patients. I love you all!

OVERCOMING
DENTAL ANXIETY

"Do not let your difficulties fill you with anxiety; after all it is only in the darkest nights that stars shine more brightly."
– Ali Ibn Abi Talib

OVERCOMING DENTAL ANXIETY

Taking care of your teeth means more than brushing and flossing. For complete care, it's important to visit a dentist every six months for a regular checkup and professional cleaning. The first step in this process is to find a dentist with whom you feel comfortable, and then schedule an appointment.

It is estimated that between 9% and 20% of Americans do not go regularly to the dentist because of anxiety or fear. That's about 30 million to 40 million people. In a survey by the British Dental Health Foundation, 36% of those who didn't see a dentist regularly said that fear was the main reason. It is indeed a universal phenomenon.

People with dental phobia have a higher risk of gum disease and early tooth loss. Avoiding the dentist may have emotional costs as well. Discolored or damaged teeth can make people self-conscious and insecure. They may smile less or keep their mouths partly closed when they speak.

Some people can become so embarrassed about how their teeth look that their personal and professional lives begin to suffer. There is often a serious loss of self-esteem. However, if we realize that oral health is part of the overall health and that going to any doctor may not be the most fun thing to do but a necessary thing to do, there are steps to take to alleviate our fear.

OVERCOMING DENTAL ANXIETY

USE THE SPACE BELOW TO DESCRIBE YOUR FEELINGS AROUND VISITING THE DENTIST.

WHAT IS
DENTAL ANXIETY???

02

"I promise you nothing is as chaotic as it seems. Nothing is worth diminishing your health. Nothing is worth poisoning yourself into stress, anxiety, and fear." - Steve Maraboli

WHAT IS DENTAL ANXIETY?????

Dental anxiety like other forms of anxiety is a feeling of worry, nervousness, or unease, typically about an imminent event or something with an uncertain outcome.

Some anxiety slowly become phobia which is a more serious condition than anxiety. It leaves people panic-stricken and terrified. Some people with dental phobia has an awareness that the fear is totally irrational but are unable to do much about it. What is even alarming is that they may be able to endure similar or more painful procedures like child bearing or tattoo graphics on their body.

They exhibit classic avoidance behavior; that is, they will do everything possible to avoid going to the dentist until it hurts so bad. People with dental phobia usually go to the dentist only when forced to do so by extreme pain.

Those experience even compounds their fear. Many are uneasy or fearful and probably has unfounded or exaggerated concerns about the appointment. Some people get to the level of pathologic anxiety or phobia which may then require psychiatric consultation in some cases.

THINK BACK TO THE FIRST TIME YOU REALIZED YOU HAD DENTAL ANXIETY. DESCRIBE THE EXPERIENCE.

WHAT CAUSES DENTAL PHOBIA AND ANXIETY?

___

03

"Whatever is going to happen will happen, whether we worry or not."
– Ana Monnar
"

WHAT CAUSES DENTAL PHOBIA AND ANXIETY?

There are many reasons why some people have dental phobia and anxiety. Some of the common reasons include:

Fear of pain. Fear of pain is one of the most common reason for avoiding the dentist. This fear usually arises from an early dental experience that was unpleasant or painful or from dental "pain and horror" stories told by others or some internet videos. Thanks to the many advances in dentistry made over the years, most of today's dental procedures are considerably less painful or even pain-free than before.

Fear of injections. Many people are terrified of needles, especially when inserted into their mouth. Beyond this fear, others fear that the anesthesia hasn't yet taken effect or wasn't a large enough dose to eliminate any pain before the dental procedure begins. Many thinks that to be numb means that they will not be able to know that their mouth exist anymore.

Fear of anesthetic side effects. Some people fear the potential side effects of anesthesia such as dizziness, feeling faint, or nausea. Others don't like the numbness or "fat lip" associated with local anesthetics. Some are even afraid of what will happen when the numbing wears off.

Feelings of helplessness and loss of control.
It's common for people to feel these emotions
considering the situation -- sitting in a dental
chair with your mouth wide open, unable to see
what's going on.

**Fear of embarrassment and loss of personal
space.** Many people feel uncomfortable about the
physical closeness of the dentist or assistants to
their face. Others may feel self-conscious about
the appearance of their teeth or possible mouth
odors.

Fear of anesthetic side effects. Some people fear
the potential side effects of anesthesia such
as dizziness, feeling faint, or nausea. Others don't
like the numbness or "fat lip" associated with
local anesthetics. Some are even afraid of what
will happen when the numbing wears off.

AFTER REVIEWING THIS SECTION, WHICH OF THE PREVIOUSLY LISTED FEELINGS HAVE CAUSED YOU TO EXPERIENCE DENTAL ANXIETY? DESCRIBE YOUR EXPERIENCES BELOW.

SIGNS OF
DENTAL ANXIETY

04

"Do not anticipate trouble, or worry about what may never happen.
Keep in the sunlight." – Benjamin Franklin

SIGNS OF DENTAL ANXIETY?

There isn't a clear boundary that separates "normal" anxiety from phobia. Everyone has fears and concerns and copes with them in different ways. However, the prospect of dental work does not need to fill you with terror. If it does, then you may need some help overcoming the fears.

Some of the signs of dental phobia include:

You feel tense or have trouble sleeping the night before a dental exam.

You get increasingly nervous while you're in the waiting room.

You feel like crying when you think of going to the dentist.

The sight of dental instruments — or of white-coated personnel in the dentist's office — increases your anxiety.

The thought of a dental visit makes you feel physically ill.

You panic or have trouble breathing when objects are placed in your mouth during a dental appointment.

If this describes you, you need to tell your dentist about your feelings, concerns and fears. He or she will help you overcome these feelings by changing the way you are treated. You also may be referred to a mental health professional.

LIST THE SIGNS OF DENTAL PHOBIA THAT YOU HAVE EXPERIENCE. DESCRIBE WHAT WAS HAPPENING IN THAT MOMENT.

_____

_____

_____

_____

_____

_____

_____

_____

_____

_____

COPING WITH DENTAL ANXIETY

05

"Trust yourself. You've survived a lot, and you'll survive whatever is coming."
– Robert Tew

COPING WITH DENTAL ANXIETY?

The key to coping with dental anxiety is to discuss your fears with your dentist. Once your dentist knows what your fears are, he or she will be better able to work with you to determine the best ways to make you less anxious and more comfortable. If your dentist doesn't take your fear seriously, find another dentist.

If lack of control is one of your main stressors, actively participating in a discussion with your dentist about your treatment can ease your tension. Ask your dentist to explain what's happening at every stage of the procedure. This way you can mentally prepare for what's to come.

Another helpful strategy is to establish a signal -- such as raising your hand -- when you want the dentist to immediately stop. Use this signal whenever you are uncomfortable, need to rinse your mouth, or simply need to catch your breath.

Other techniques that might help include:

1. Speak up
Anyone with anxiety knows sharing your feelings makes a world of difference. If you're tense or anxious, do yourself a favor and get your concerns off your chest. Your dentist and dental team are better able to treat you if they know your needs.

Tell your dentist about your anxiety. When you book your appointment, tell the receptionist you're nervous about dental visits. Remind the dentist and dental staff about your anxiety when you arrive. Share any bad experiences you may have had in the past, and ask for suggestions on coping strategies.

Don't be afraid to ask questions. Sometimes knowing what is going to happen alleviates any fears of the unknown.

Agree on a signal. Let your dentist know by raising your hand if you need to take a break during an exam.

If you experience pain even with a local anesthetic, tell your dentist. Some patients get embarrassed about their pain tolerance or don't want to interrupt a dentist during a procedure. Talk with your dentist about pain before it starts so your dentist knows how to communicate with you and make it more comfortable.

2. Distract yourself
Taking your mind off the exam may seem impossible when you're nervous, but there are some things that that can help distract your thoughts.

Wear headphones. If the sound of the drill bothers you, bring headphones so you can listen to your favorite music or audiobook. Some dental offices even have televisions or show DVDs.

Occupy your hands by squeezing a stress ball or playing with a small handheld object, like a fidget spinner.

Imagine your happy place and visualize yourself at a relaxing beach or garden.

3. Use mindfulness techniques
Relaxation starts in the mind. Try deep breathing exercises to help relax tension in your muscles.

Count your breaths. Inhale slowly and then exhale for the same number of counts. Do this five times while you're waiting for your appointment, or during breaks while you're sitting in the dental chair.

Do a body scan. Concentrate on relaxing your muscles, one body part at a time. Start with your head and work your way down to your toes. For example, you can focus on releasing tension starting in your forehead, then your cheeks, your neck and down the rest of your body.

It's common for patients to feel dental anxiety or uneasiness before or during their dental appointments.

Here are some tips to help you overcome your fear of the dentist:

Talk with your dentist and/or dental hygienist about your dental anxiety so they can explain what the procedure really entails and you know what to expect. They're there to make your experience as comfortable and stress-free as possible.

Patients often say that their fears were worse than the actual procedure and that if they'd only known what to expect they would have been less anxious.

Wear headphones during your appointment to help block out sounds. Be sure to discuss this with your dentist or dental hygienist prior to your appointment.

Request a numbing agent. Your dental hygienist and dentist can often apply a numbing gel or use a local anesthesia to minimize pain.

Everyone's pain threshold is different. Don't be afraid to ask for a numbing agent if you weren't offered one or if you're in pain—your dentist will not want to proceed unless you're comfortable.

Ask about anti-anxiety medication. Some practices offer patients anti-anxiety medication to help reduce dental anxiety during the appointment. Please ask your dentist before the day of your visit about anti-anxiety medications.

Use the chair-side TVs to watch a favorite show during your procedure. Your dentist can also use the monitor to show you your x-rays and to help you to better understand your procedure. TVs are available in most exam rooms.

Bring someone you trust to your appointment to help provide comfort and reduce dental anxiety.

Arrive early to your appointment. Stress management can start just by reading a magazine in the waiting room.

LIVING WITH DENTAL ANXIETY CAN BE EXTREMELY DIFFICULT. IDENTIFY WHICH COPING METHODS YOU WILL USE GOING FORWARD.

_____

_____

_____

_____

_____

_____

_____

_____

_____

Journal
Space

JOURNAL

JOURNAL

JOURNAL

JOURNAL

JOURNAL

JOURNAL

JOURNAL

JOURNAL

JOURNAL

JOURNAL

JOURNAL

JOURNAL

_____

_____

_____

_____

_____

_____

_____

_____

_____

_____

www.ingramcontent.com/pod-product-compliance
Lightning Source LLC
Chambersburg PA
CBHW021049180526
45163CB00005B/2350

* 9 7 8 0 3 5 9 7 0 5 7 5 7 *